Technology
A Closer Look

Macmillan/McGraw-Hill

Content Consultants

Robert L. Kolenda
Science Coordinator, K-12
Neshaminy School District
Langhorne, PA

Paul Mulle
Science Supervisor
Camden City Schools
Camden, NJ

Patricia Cavanagh
Elementary School Teacher
Merrimac Elementary School
Holbrook, NY

Pat Marengo
Sixth Grade Teacher/
Technology Specialist
District 58 Elementary School
Downers Grove, IL

RFB&D
learning through listening

Students with print disabilities may be eligible to obtain an accessible, audio version of the pupil edition of this textbook. Please call Recording for the Blind & Dyslexic at 1-800-221-4792 for complete information.

Internet Disclaimer
Visit www.macmillanmh.com to learn more about technology. You will also find links to other Web sites. These Web sites are not run by Macmillan/McGraw-Hill. When using the Internet, be safe and protect your privacy. Make sure a teacher, parent, or guardian is around. Never tell someone your full name, address, passwords, or other personal information. Do not respond to e-mails or messages from strangers.

The McGraw·Hill Companies

Macmillan McGraw-Hill

Send all inquiries to:
Macmillan/McGraw-Hill
8787 Orion Place
Columbus, OH 43240-4027

Printed in the United States of America

ISBN-13: 978-0-02-285949-7
ISBN-10: 0-02-285949-7

4 5 6 7 8 9 073/073 11 10 09

Contents

Lesson 1

Main Idea
People have always used technology, which is how we change nature to meet our needs.

Vocabulary
technology, p. 2
natural resources, p. 2
system, p. 3
goods, p. 4
services, p. 4

What Is Technology?

When hearing the word technology, most people think of things like computers, cell phones, and satellites. But it's more than that. **Technology** is the way we adapt, or change, nature to meet our needs. It is all the tools we design, make, and use to solve problems. Technology dates back to when the first human picked up a rock and used it as a tool!

Rocks are a natural resource. A **natural resource** is a material from Earth that people use to meet a need. Everything you see is either a natural resource, or something designed using natural resources.

Some of the earliest wheels were simple wooden disks.

Transportation Systems

Technology and Systems

People use science to design and build systems. A **system** is a group of parts that work together to do something. Systems help people move, communicate, and build things.

Transportation systems move people (and things) from one place to another. Long ago, people had to walk or ride horses to get anywhere. The invention of the wheel changed that. Later, the engine was invented and the automobile came to be. Today, transportation systems include vehicles, roads, bridges, and water routes.

A horse-drawn bus

Quick Check

Summarize How has technology changed the way people travel?

Read a Picture

What transportation systems can you find in this city? Do you have the same systems where you live?

Clue: Make a three-column chart with the headings LAND • WATER • AIR.

Power and Production

Goods are objects people make and use. Where do we get the goods we need and want? At first people made most goods themselves. Then some began to make extra goods to sell. People also supplied **services**, or performed work for money.

In time, small businesses started to grow. They made and sold goods. Some made and sold furniture; others made and sold bread or clothing. Customers bought what they needed.

▲ **Before the 1830's shoes were made by hand.**

▲ **Hand-made shoes were sold in this shoe store in the early 1900's.**

Later new inventions and materials, like machines and steel, led people to build factories. This allowed them to make a lot of items at one time. Many workers were hired to run machines. Goods were transported all over the world.

Quick Check

Think About It **How has technology changed the way you get the things you need?**

Most of today's shoes are made by machines in factories.

Tech Activity

Design a Boat

Design a boat to transport a cargo of marbles across water.

Materials **aluminum foil, newspaper, large container filled with water, jar of marbles, scale, small plastic bag**

What to Do

1. Fold foil into boat shape.
2. Put the boat in water.
3. Add marbles until the boat sinks. Count the number of marbles and put them in a small plastic bag.
4. Weigh the bag of marbles. Record its weight.
5. Refine the design of your boat to increase its capacity. Load the newly designed boat with the marbles you weighed. Did the boat stay afloat?
6. Compare with classmates. Did your boats differ in design? Which boat held the most weight?

Explore More

Build a bridge and cable system using simple machines (like rulers as inclines, and a spool of thread as a pulley). Write a procedure that describes how to pull your cargo up and over the bridge.

Are You There Yet?

Have members of your family ever gotten lost while driving? Well, they don't have to anymore, thanks to GPS (Global Positioning System). GPS is made up of 24 solar-powered satellites in orbit about 12,000 miles above Earth. The satellites orbit Earth twice a day. (Tiny rockets keep each satellite in the right orbit.) The satellites transmit signals to receivers on Earth.

Some cars have a GPS receiver. It uses the satellite signals to figure out the car's location and displays it on a map. Once the receiver knows the car's location, it tells the driver how to get to his or her destination.

GPS signals can pass through clouds and glass, but not solid objects. So sometimes tall trees or buildings throw GPS data off by a few meters. But that's a lot better than being totally lost!

Write About It

Newspaper Article Research and write a newspaper article about GPS technology. What are the latest advances? How will they affect the future?

LOG ON e-Journal Research and write about it online at www.macmillanmh.com

Turn right!
Now turn left.
You have arrived.

Lesson Review

Think, Talk, and Write

Write your answers on a separate sheet of paper.

1. **Main Idea** How are technology and nature connected?

2. **Critical Thinking** How has technology changed people's lives over the years?

3. **Complete the Sentences** Choose your answer from the following words.

Word Box	
roads	systems
adapt	bridges
waterways	needs
service	Global Positioning System

A Technology is how people ________, or change nature to meet our ________.

B People use natural resources to design and build ________.

C Today's transportation systems include ________, ________, and ________.

D People who do work for others supply a ________.

E The letters GPS stand for ________.

4. **Read a Photo** What goods do you see in the photo above? What are some services?

5. **Writing Link** Turn to the front of your book and skim the Table of Contents. Then browse the rest of the book. Predict what you'll learn from this book. Which part do you think will be the most interesting? Why?

LOG ON e-Tech Link For more on technology is go to www.macmillanmh.com

Lesson 2

Main Idea

People follow steps in a design process to turn ideas into inventions.

Vocabulary

patent, p. 9

design process, p. 9

model p. 12

prototype, p. 12

Ideas and Inventions

Think of an invention you use every day. It began as an idea in someone's mind. Inventions are the result of a long process that starts with a need and leads to an idea.

Inventions

Thomas Edison invented many things, but the light bulb made him famous. Edison saw a need. He thought about making a light bulb that people could afford. One important challenge was finding a material to use as a filament. A filament is a thin material that heat passes through to produce a glow. After spending more than two years searching for the right material, Edison tried bamboo. The bamboo filament glowed for days!

◀ Thomas Edison's original light bulb

Thomas Edison working in his laboratory. ▶

When Margaret Knight was growing up, paper bags were designed to be like big envelopes. They could not stand on their own. Knight saw a need for a bag that could stand up. She invented a machine that made bags with flat bottoms.

In 1870, Margaret Knight became the first woman to receive a **patent**. A patent gives a person the right to claim an invention as their own. Knight started her own paper bag company that same year!

The Design Process

Edison and Knight both took certain steps to bring about their ideas. These steps are called the **design process**.

Some inventors make existing objects work better. Others find new ways to use objects. Still others come up with ideas for new things. No matter what they invent, all inventors use the design process. On the next few pages, you will learn how this process works.

▲ **Margaret Knight's paper bag machine**

Quick Check

Sequence Before coming up with their ideas, what was the first thing both Edison and Knight noticed?

Steps in the Design Process

1 Identify a Problem

The first step in the design process is to identify a problem. A problem can be a need or a want that if solved might make life easier, better, or just more fun. How do you identify a problem? You observe people and patterns. You ask questions.

Suppose your pet's toy rolls behind a large piece of furniture. The furniture is too heavy to move, and you can't reach far enough underneath or behind it to grab the toy. Your pet sure wants that toy!

You have just identified a problem. You need a safe way to get the toy back!

What's the Problem?

Read a Picture

What need can you identify from this picture?

Clue: Explain the problem in your own words.

2 Propose a Solution

The next step is to think of possible solutions and sketch them. Solutions are ideas to solve the problem. What are some ideas that could help solve your problem?

▲ You could put something sticky at the end of a meter stick and try to get the object to adhere to it.

▲ You could use a long set of tongs to pick up the object.

After sketching your ideas, pick the best one to try. Ask questions to figure out the challenges in each idea. What materials are needed? Are they available? Are there any risks involved? How much would it cost to make?

To answer these questions, talk to people, make observations, and do research.

▲ You could use a meter stick with a magnet on a string to attract the object.

Quick Check

Sequence Which do you do first, sketch an idea, or identify the problem?

③ Build a Model

Build a **model** to test your ideas. A model helps you understand how your solution works. Sometimes you need to build a **prototype** of your invention. A prototype is a life-size working model that can be tested.

Building a prototype of the "tong" solution

④ Test the Design

Inventors test their designs by asking questions. Does the invention do what it was designed to do? Why? Why not? A smart inventor also asks other people to test the invention and then listens to their feedback.

Say the invention doesn't work. In this case, the inventor would have to rethink the design. Rethinking could lead to a completely different solution. When a design works, inventors often improve or refine it until they're satisfied the invention really works. Next they might try to sell their design to a manufacturing company. An important part of this step is knowing how much it will cost to make the product.

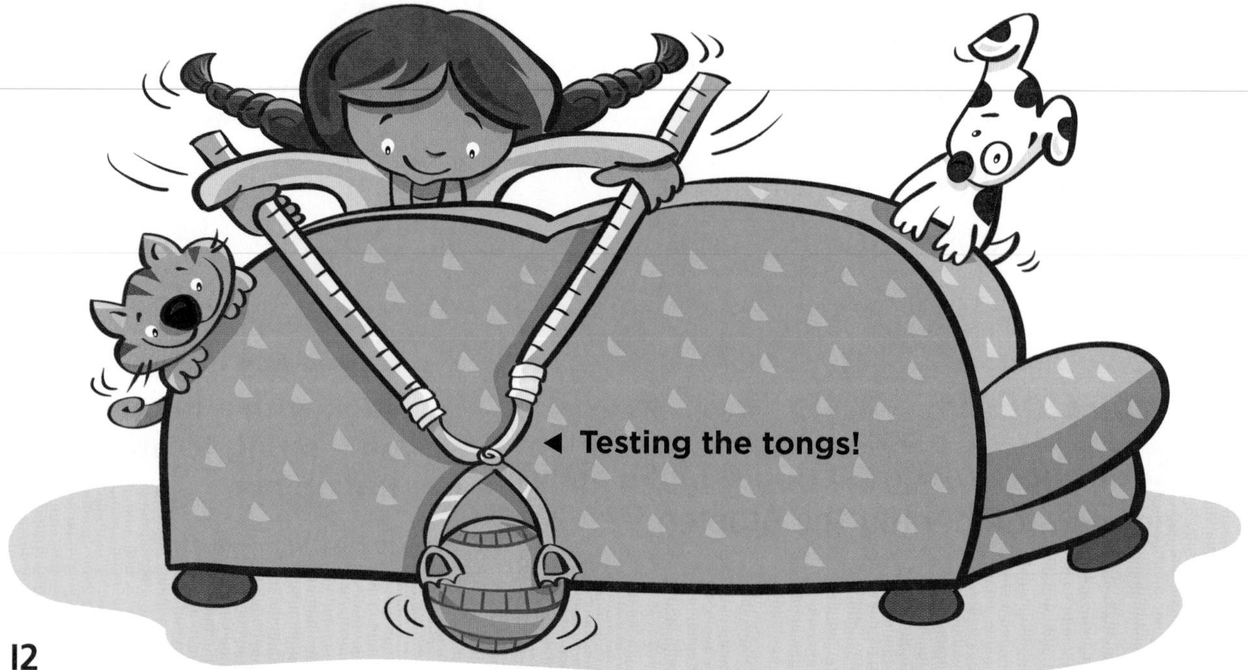

◄ **Testing the tongs!**

5 Explain the Invention

Finally, you communicate how you solved the problem. Communication can be like a show and tell. You can also use group discussions, written reports, and pictures. If you've made a product you want to sell, you name it and advertise it.

▼ **Communication is the last step in the design process.**

1 identify a problem
2 propose a solution
3 build a model
4 test the design
revise the design
5 explain the invention

Tech Activity

Use the Design Process

Materials **See page 11.**

1. Build a prototype of each design solution.
2. Test each prototype to see how well it picks up everyday objects.
3. Record your results in the worksheet for this activity.

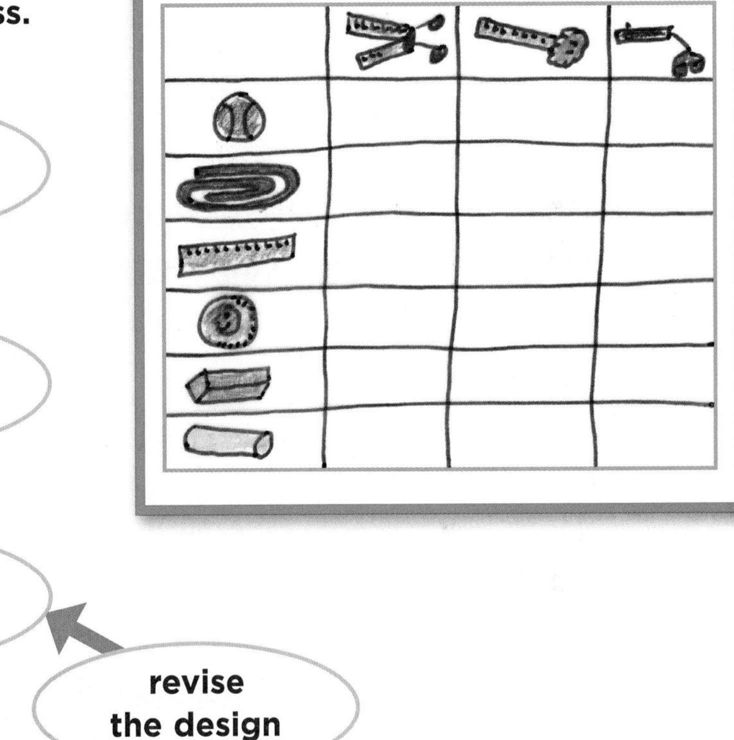

Quick Check

Sequence When in the design process should you set a price for your invention? Why?

Designed for Speed

Athletes have a need for speed. But air or water can slow them down. Air and water create *drag*, a type of friction.

What is the solution? Scientists work to design equipment and clothing that reduces drag. Such designs are referred to as *aerodynamic* (AYR•oh•dy•NAHM•ik).

Aerodynamic body suits fit like a second skin. The coverings of elbows, knees, and other joints are designed to prevent creasing. Creases can trap air and slow an athlete's speed. Many Olympic competitors use such sportswear.

Designers of aerodynamic suits test them. They may place a model in a wind tunnel to see how air moves over the body. Then they use this information to improve their designs and fabrics.

▲ **Speedskaters use aerodynamic body suits and helmets.**

◀ **Wind tunnels like this help designers understand the problem of drag.**

Write About It

Expository Nonfiction

Research other aerodynamic objects, like cars and airplanes. How are they designed to cut down on drag?

LOG ON e-Journal Research and write about it online at www.macmillanmh.com

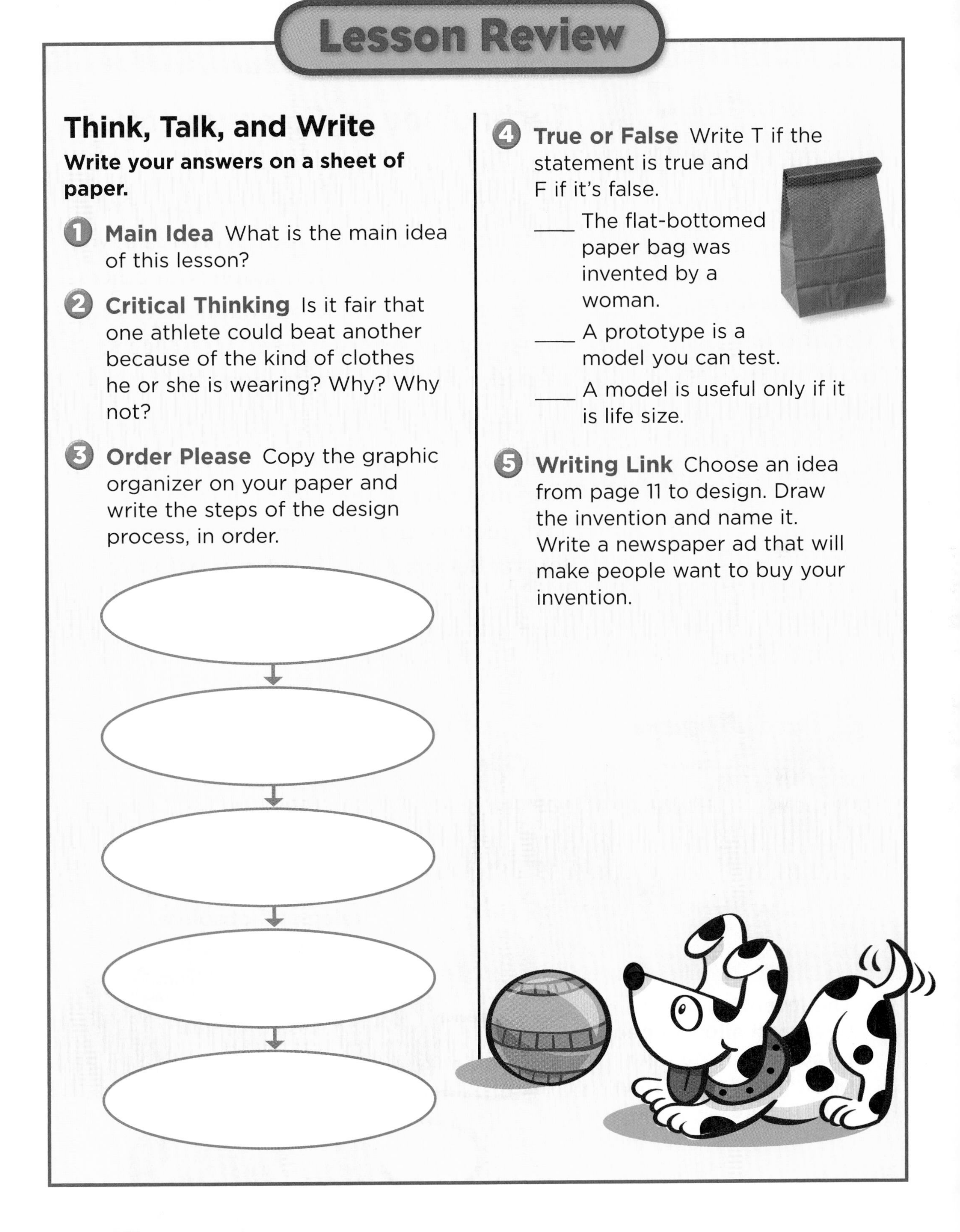

Lesson Review

Think, Talk, and Write

Write your answers on a sheet of paper.

1. **Main Idea** What is the main idea of this lesson?

2. **Critical Thinking** Is it fair that one athlete could beat another because of the kind of clothes he or she is wearing? Why? Why not?

3. **Order Please** Copy the graphic organizer on your paper and write the steps of the design process, in order.

4. **True or False** Write T if the statement is true and F if it's false.

   ____ The flat-bottomed paper bag was invented by a woman.

   ____ A prototype is a model you can test.

   ____ A model is useful only if it is life size.

5. **Writing Link** Choose an idea from page 11 to design. Draw the invention and name it. Write a newspaper ad that will make people want to buy your invention.

Lesson 3

Main Idea

Technology has changed and improved the way we communicate.

Vocabulary

communicate, p. 16

telegraph, p. 16

Technology in Communications

Do you use the telephone? Do you write letters or send e-mail? These are all forms of communication. When you **communicate**, you exchange ideas and information with others. Long ago some people communicated using smoke signals or drumbeats. Messengers carried mail by foot or on horseback.

With the discovery of electricity, new forms of communication were possible. The invention of the telegraph in the 19th century changed communication forever. **Telegraph** machines sent patterns of long and short clicks over electric wires. The series of clicks—called Morse Code—spelled out words.

▼ **The invention of the telephone made the telegraph obsolete.**

▲ **The invention of the telegraph allowed people to communicate over greater distances than ever before.**

The telephone soon replaced the telegraph. Early phones needed operators to connect callers. Later phones had dials.

Today cell phones are everywhere. Instead of sending signals over wires, cell phones use radio waves. Most cell phones can also send text messages, pictures, and videos.

Some phones use satellites to communicate information. A *satellite* orbits Earth.

Quick Check

Summarize How has communication changed over the last 200 years?

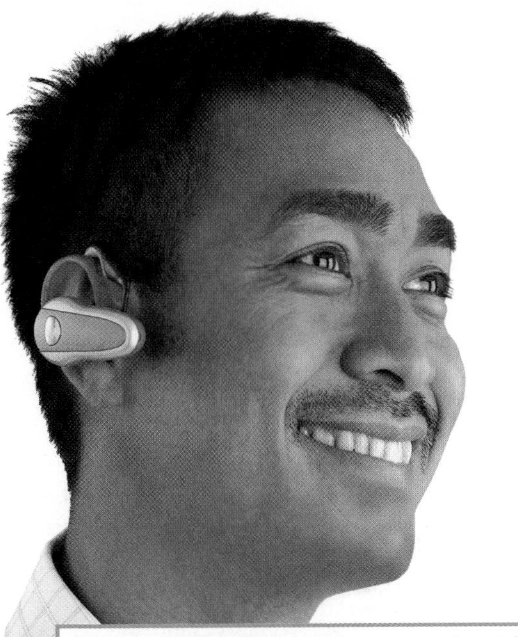

Some modern communication tools are wireless.

Tech Activity

Morse Code Message

Materials **electric circuit (battery, wires, bulb), Morse Code chart**

1. Work with a partner to build a simple electric circuit. Test the switch. ⚠ **Be Careful.** The wires may get warm.
2. Think of a short, simple message to send to your partner. Write your message in dots and dashes using the chart shown.

Letter	Code	Letter	Code	Letter	Code
A	.-	J	.---	S	...
B	-...	K	-.-	T	-
C	-.-.	L	.-..	U	..-
D	-..	M	--	V	...-
E	.	N	-.	W	.--
F	..-.	O	---	X	-..-
G	--.	P	.--.	Y	-.--
H	....	Q	--.-	Z	--..
I	..	R	.-.		

3. Send your message by flashing the light once for a dot or three times quickly for a dash. Count 3 seconds between each letter. Count 5 seconds between each word.
4. As you flash your message, have your partner write the pattern in dots and dashes. Your partner can then use the chart to decode your message.

Draw Conclusions

In what kind of situations would Morse Code be useful? Why don't we use this technology today? How do people communicate instead?

Communication Systems Connect People

Think about the transportation system discussed in Lesson 1. It involved more than just vehicles. Well, a communication system involves more than just phones.

A communication system has four basic parts—input, process, output, and feedback. Let's say you write a letter inviting a friend to visit. The letter is the *input*. The *process* is how you send it—by mail. The delivery of the letter to your friend is the *output*. A return letter from your friend is the *feedback*.

A communication system also includes the equipment, machines, and people needed to run it. That means mailboxes, scales, post offices, and mail carriers are part of the postal system.

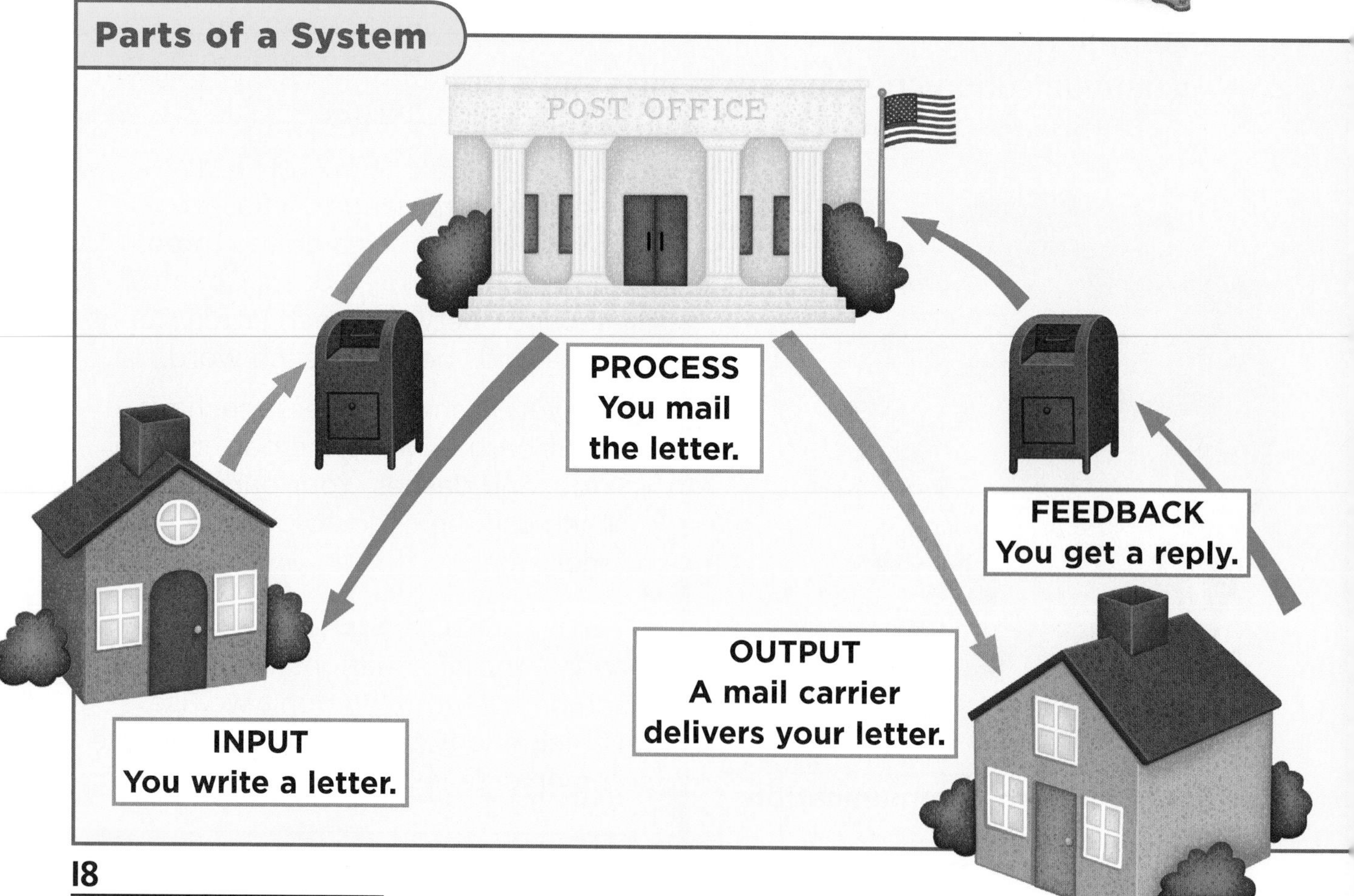

The Role of Computers

A computer can be a communication system. When you hit a key, or click a mouse, you are inputting data. The computer processes the data and you see the output on the monitor. You can adjust the output. That is the feedback.

Computers were invented to do math. The first one was bigger than an elephant. Over the years, computers got smaller, faster, and smarter. Today we use computers to write, store, and send information. We use them to browse the Internet, do research, send e-mails, prepare presentations, and even make movies!

Quick Check

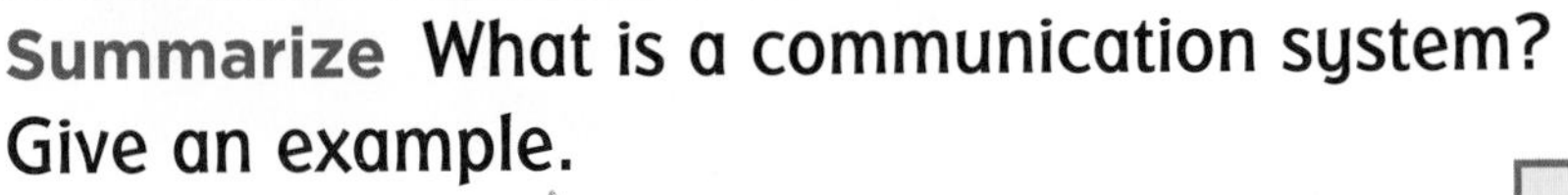

Summarize What is a communication system? Give an example.

Read a Diagram

What would happen if this system had no output?

Clue: Follow the arrows and read the captions.

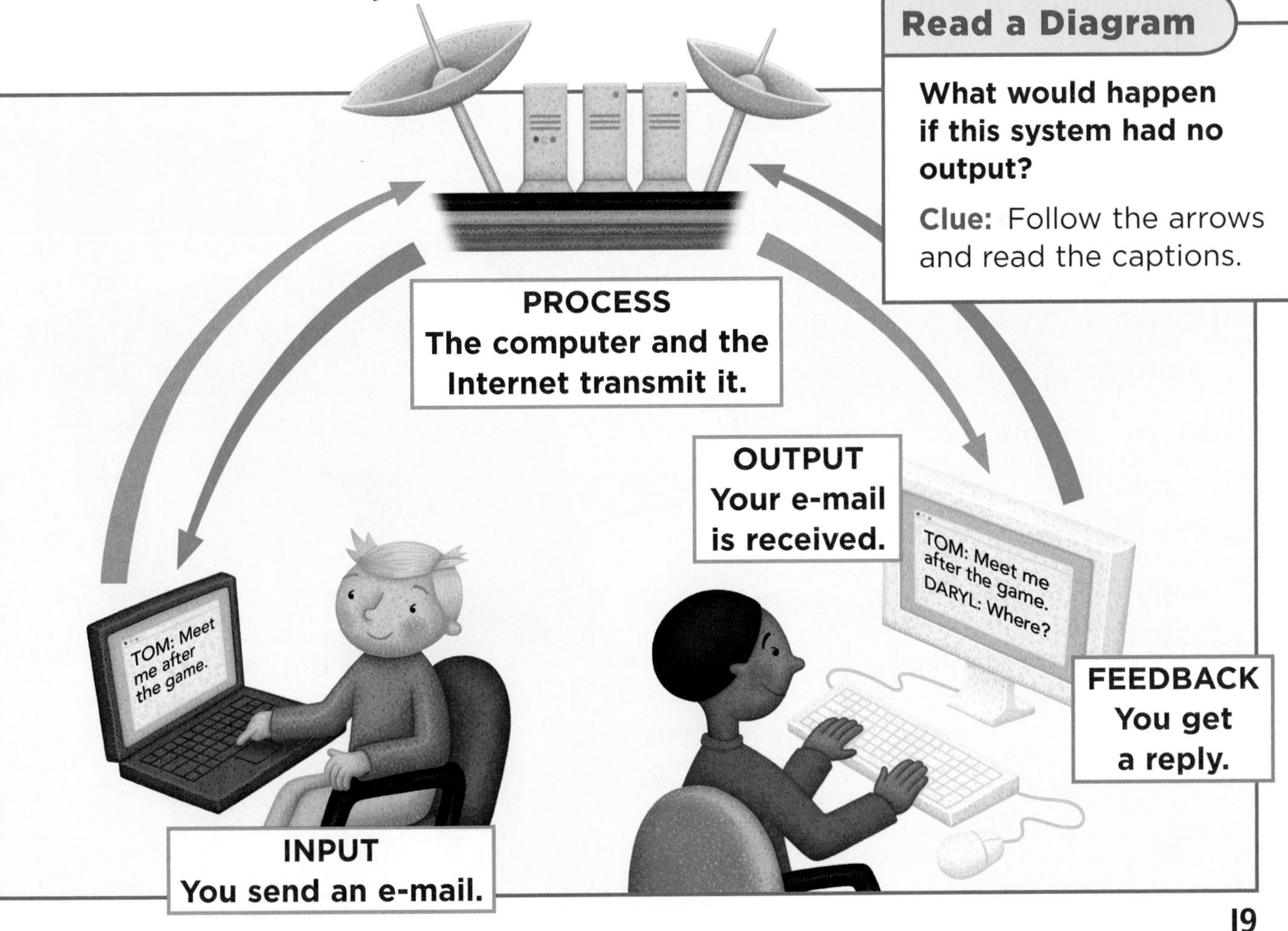

Communicating Using Images

People communicate with images as well as with sounds and letters. You see photographs every day in magazines, on the Internet, and in this book.

Cameras produce images, or photos. Early cameras used film, a material that is sensitive to light. When light enters the camera and hits the film inside, it makes an image.

In order to see the image on the film, you must treat, or develop, the film with chemicals in a dark room. In 1948, an American named Edwin Land invented the instant camera. Instant cameras push the picture out, and you can watch the image develop!

▲ **nineteenth-century camera**

You are probably more familiar with *digital cameras*. They don't use film. They have tiny sensors that turn light into electrical energy. The camera sends the image to a memory card. The card stores the image and a computer reads it.

Many cell phones have digital cameras built in. Pictures can be e-mailed to friends and family right away.

digital image from camera

Images that Move!

The invention of the camera paved the way for the first movies, or motion pictures. The first movies were black and white "silent films." You could see actors' lips move, but you couldn't hear what they said! In 1910, Thomas Edison made the first talking picture. Soon the "talkies" took over.

In the early days of film, all stunts had to be performed by the actors. There were no special effects.

Movies have come a long way since those old, silent films in black and white. Today's technology lets filmmakers add color, sound, special effects, and computer animations. Some movie scenes look so real, you have to be reminded that you're watching technology in action!

Quick Check

Summarize How is technology used to communicate through pictures?

Today, movie makers use computer-made animations and special effects.

Transmitting TV

Today television is a big part of how we communicate. Cable and satellite systems offer hundreds of channels to choose from. You can watch sports events live. You can check the news and the weather forecast any time of day.

Have you ever wondered how TV works? First radio waves are transmitted, or broadcast, into the air. Satellites pick up these radio waves, or signals, and relay them over long distances.

TV antennas or satellite disks receive the waves. Sensors in the TV decode the signals. When the TV is turned on, the screen shows images. You watch the images to learn or be entertained.

satellite

transmitter

television set ▶

Write About It

Fantasy Write a story from the point of view of a newscast traveling through the air to your TV. What happens along the way?

LOG ON e-**Journal** Research and write about it online at **www.macmillanmh.com**

Lesson Review

Think, Talk, and Write

Write your answers on a separate sheet of paper.

1 Main Idea and Supporting Details What are some ways we use technology in communications?

2 Critical Thinking Describe TV transmission as a system with input, process, and output.

3 Complete the Sentences Choose your answer from the following words.

Word Box	
electrical energy	ideas
radio waves	information
electric wires	exchange
film	sensors

A Communication is the ________ of ________ and information with others.

B Telegraph machines send patterns of short and long clicks over ________ ________.

C Instead of transmitting signals over wires, cell phones use ________ ________.

D Early cameras used ________ on which the image formed.

E Digital cameras have tiny ______ that turn light into ________.

4 Word Scramble Unscramble each word. Use the words in the Word Box to help.

A TIPUN
B SANITTRM
C MACIOTUNICNOM
D MAREAC
E. RPTOCEUM

Word Box	
communication	transmit
input	computer
camera	

5 Read a Diagram Label the input, process, output, and feedback in the illustration below.

A ______________________

B ______________________

C ______________________

D ______________________

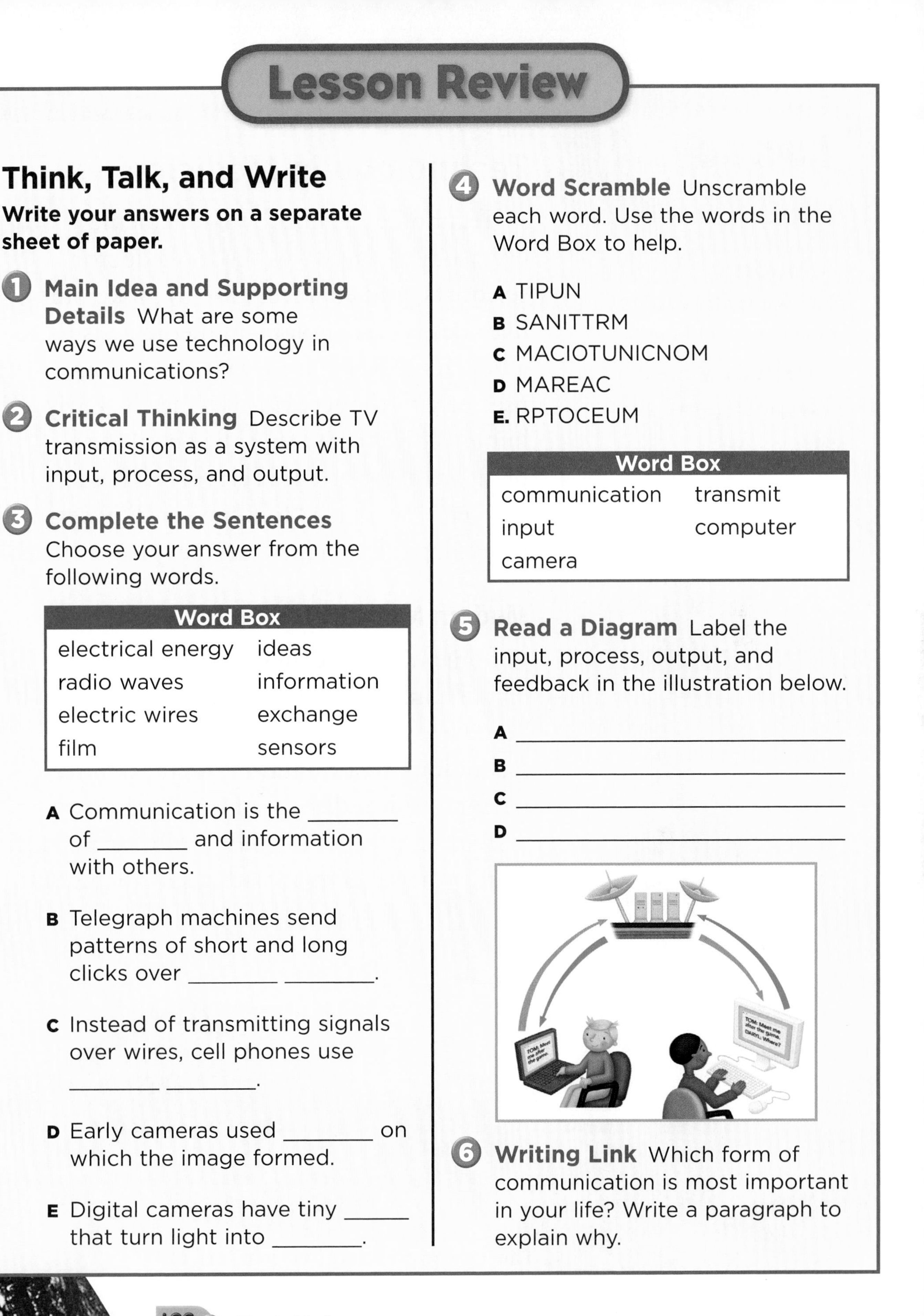

6 Writing Link Which form of communication is most important in your life? Write a paragraph to explain why.

LOG ON e-Tech Link For more on technology and communication go to www.macmillanmh.com

Lesson 4

Main Idea
Technology has improved health and medicine.

Vocabulary
biotechnology, p. 29
traits, p. 29

Technology in Medicine

Technology is improving medicine every day. It improves the tools doctors, scientists, and engineers use to make our lives better. Technology in medicine isn't a new idea. Remember that technology uses nature to meet peoples' needs. Early humans discovered that they could use plants to treat or cure certain sicknesses. For centuries, such "folk medicine" was a technology that helped sick people get better.

Modern Medical Technology

Some modern drugs are also based on substances that come from plants. To develop these medicines, science and technology work hand-in-hand. Asthma (AZ•muh) is a disease of the respiratory system. It affects the ability to breathe. How could you develop an asthma medicine?

The common foxglove is used to make Digoxin, medicine that treats heart disease.

First you would need to know what causes asthma. You would also need to know how different medicines and other substances affect the body. Then you would form a hypothesis about what substances to put in your new medicine. You would test the medicine to see how well it worked.

Quick Check

Cause and Effect How does the discovery of plant medicines come about?

A Medical Technology System

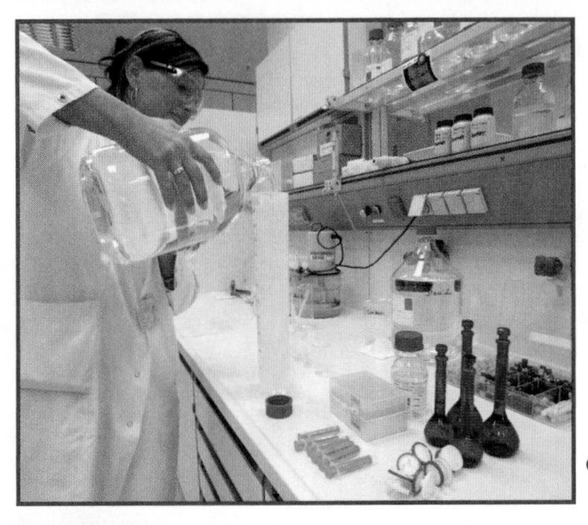

INPUT People who work in a lab input the chemicals and/or plant parts that make up the medicine.

PROCESS Machines mix, or process, the ingredients.

OUTPUT The output is medicine in bottles or other containers. Doctors prescribe the medicine to patients.

FEEDBACK Feedback comes from the people who try the medicine. Patients tell the doctor if the medicine helps their asthma.

More than Medicine

Medical technology is more than just making medicines. It includes developing tools and methods to help people stay healthy and safe. Here, too, science and technology work together. A stethoscope (STETH•uh•skohp) helps doctors hear inside your chest. This tool couldn't have been invented unless someone knew how sound travels!

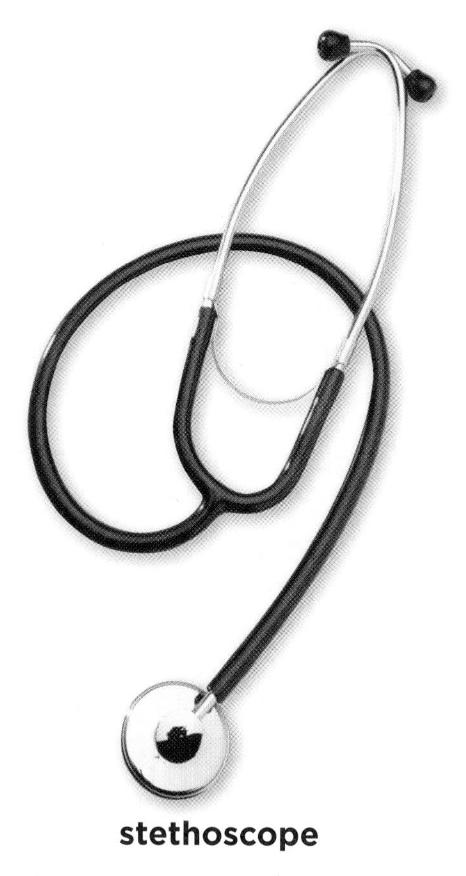
stethoscope

In 1895, a German scientist discovered a light ray that passed through some solid objects more than others. This discovery led scientists to develop the first X-ray machine. The machine lets doctors take a picture of the inside of the body. X-rays help doctors detect broken bones, cavities in teeth, and some lung problems.

Modern Medical Tools

Today, doctors still use X-rays, but they have other tools to help them see inside your body. Complex machines like CAT scanners, MRIs, and ultrasound help them find problems.

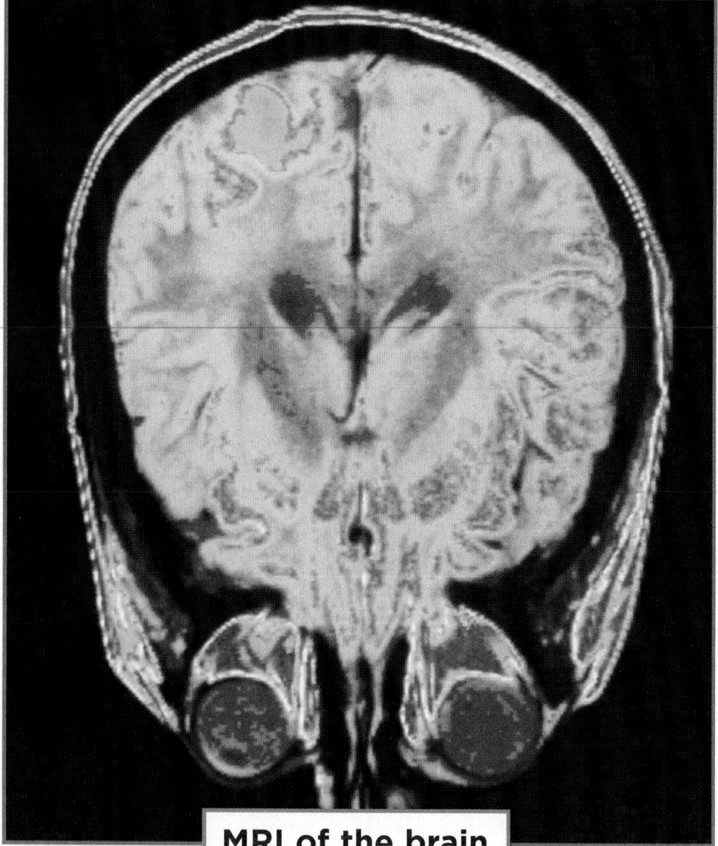
MRI of the brain

Other tools help doctors go inside the body to look around and fix things. An endoscope is such a tool. It's tiny camera with a light attached to it helps doctors see inside an organ or body cavity. This tool helps doctors diagnose illnesses.

Newer medical technologies include robots and lasers. These help doctors operate on patients. Doctors can operate using tiny robots inside the body. Lasers can cut through the skin or parts of the eye without causing bleeding.

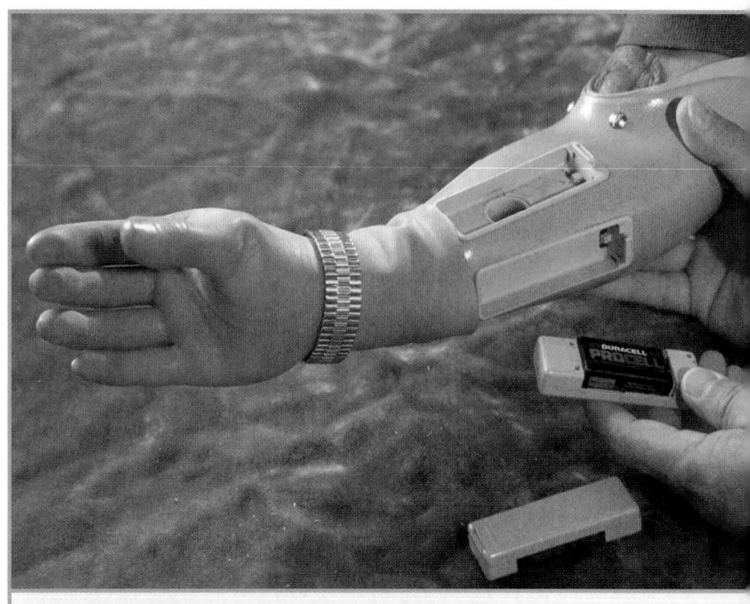

An artificial arm can perform the functions of a real arm.

Medical technology can help people who have lost parts of their body. Today artifical body parts look, bend, and work like real ones.

Quick Check

Cause and Effect How might X-rays be used to help you stay healthy?

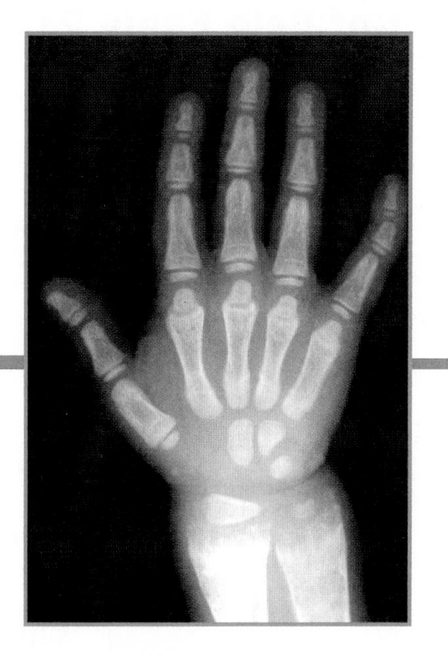

Tech Activity

Model an X-ray

Materials **goggles, disposable glove, white paper, spray bottle with mixture of water and tempera paint, dark crayon or marker**

What to Do

1. Put on the glove and goggles.
2. Place your hand over the paper.
3. Spray the liquid over your hand.
4. **Observe** Look at the print you made. Does it look like your hand?
5. **Infer** Draw a line across one of the fingers. If this were an X-ray, what would the line mean?

A Healthy Environment

Technology has solved many problems, but it has also created some. DDT is an example of a problem created by technology. DDT is a chemical that kills weeds and other pests. At one time many farmers sprayed their crops with DDT. This poisoned water sources. Birds that had eaten poisoned fish laid eggs that didn't hatch. The bald eagle—our national symbol—almost disappeared as a result. DDT caused problems for other wildlife as well.

In 1972, the government outlawed the use of DDT. Fish and wildlife began to recover. Today the population of bald eagles is healthy.

The story of DDT taught scientists to solve problems in ways that won't hurt the environment.

Number of Bald Eagle Pairs in 48 States

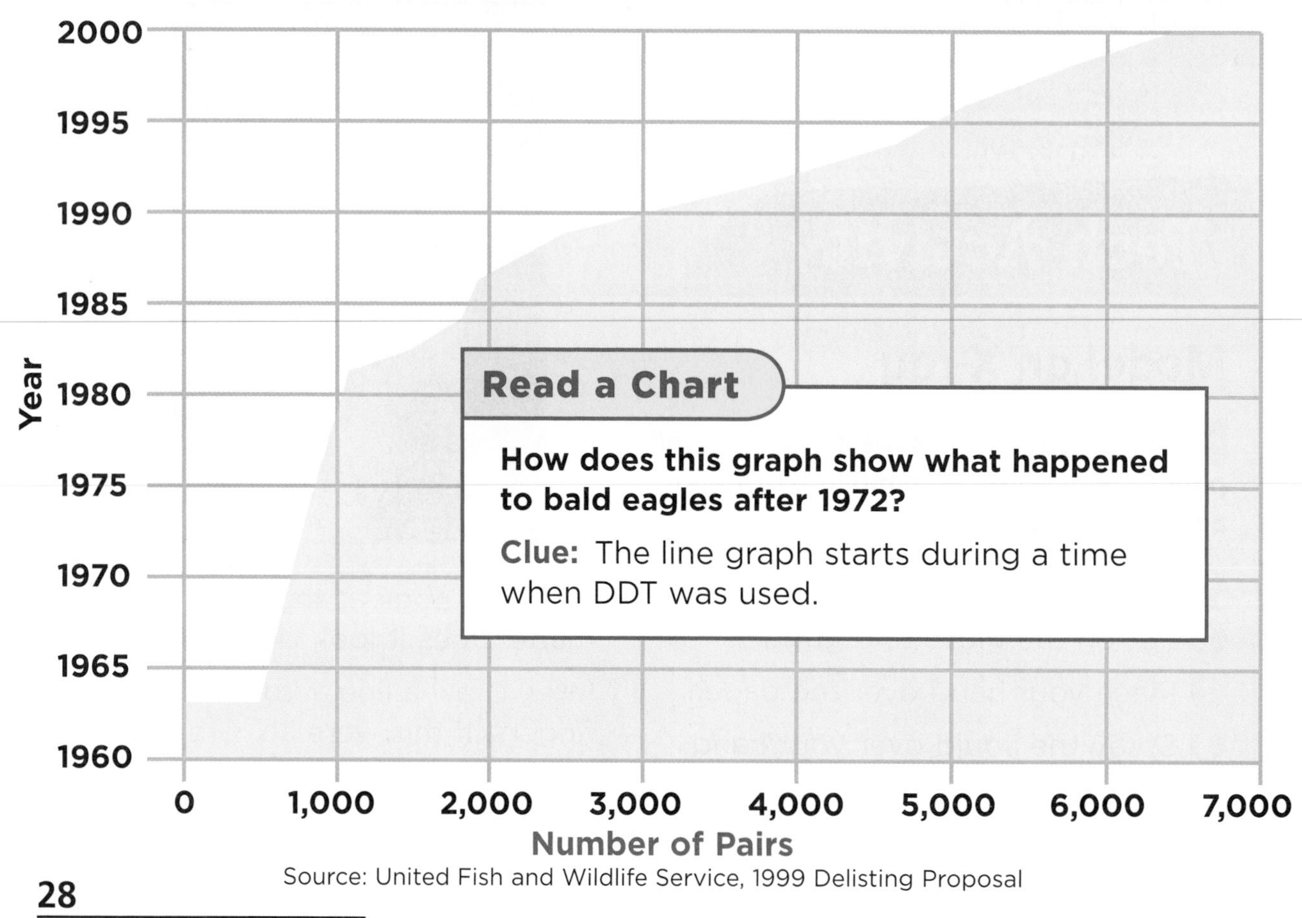

Source: United Fish and Wildlife Service, 1999 Delisting Proposal

Read a Chart

How does this graph show what happened to bald eagles after 1972?

Clue: The line graph starts during a time when DDT was used.

Ladybugs eat aphids, which benefits plants.

Biotechnology

You read on page 24 that the foxglove plant is used to make medicine. This is just one example of **biotechnology**. Biotechnology uses living things or living systems to meet human needs.

Another example of biotechnology is pest control. Ladybugs can be very helpful in a garden. Ladybugs eat aphids. Aphids can damage fruit trees and vegetable crops. Many farmers and gardeners use ladybugs to keep their plants healthy!

One area of biotechnology involves mixing characteristics of living things. This field is called genetic engineering. Genetic engineering allows scientists to create organisms with different traits.

Some plants have traits that help them resist certain pests. These traits can be combined with the traits of other plants, making them pest-resistant, too.

Scientists hope biotechnology can help solve human problems without harming other living things and their environments.

Quick Check

Cause and Effect How can we use biotechnology to grow better crops?

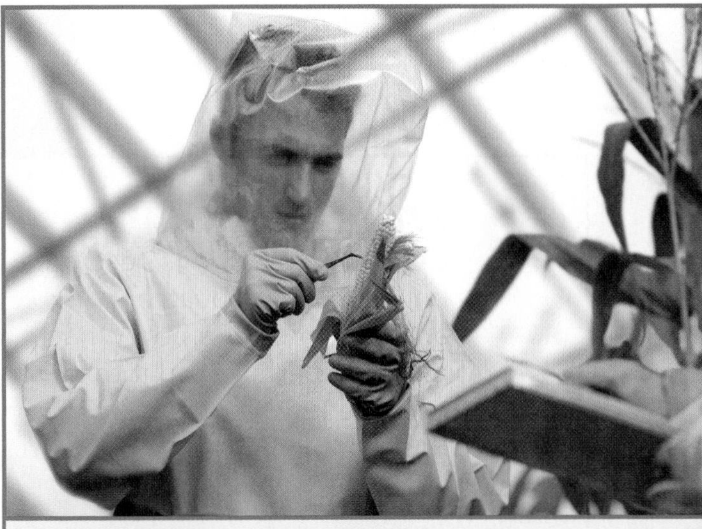

A researcher experiments with new ways to grow corn.

Not Just for Humans

Medical technology doesn't only help people. Animals can have medical problems, too. A veterinarian's office has important technology for keeping dogs, cats, and other pets healthy. Veterinarians offer medical tests, X rays, medicines, and vaccinations. What if an animal needs an operation? The vet may operate using the same kinds of tools and methods used on humans.

Medical technology helps animals in the wild, too. In the past, an animal that hurt or lost a limb was often killed. People thought there was nothing else they could do. Without the use of its limb, the animal couldn't survive in its environment. This is no longer the case. Mammals, such as horses, can get artificial knees or other joints. From kiwi birds to elephants, artificial limbs help many lives!

Write About It

Descriptive Writing Describe how you would build an artificial limb for a pet. What different parts would it take to make it work? Draw a design of your idea.

LOG ON **e-Journal** Research and write about it online at **www.macmillanmh.com**

Lesson Review

Think, Talk, and Write

Write your answers on a sheet of paper.

1. **Main Idea** Discuss some ways that medical technology has made life better or easier.

2. **Critical Thinking** Chemicals can kill insects that cause diseases in humans. For instance, mosquitoes can spread West Nile virus. People use chemicals to control mosquitoes and prevent the disease in humans. However, such chemicals can also harm humans. How would you decide whether to use chemicals to control pests in your area?

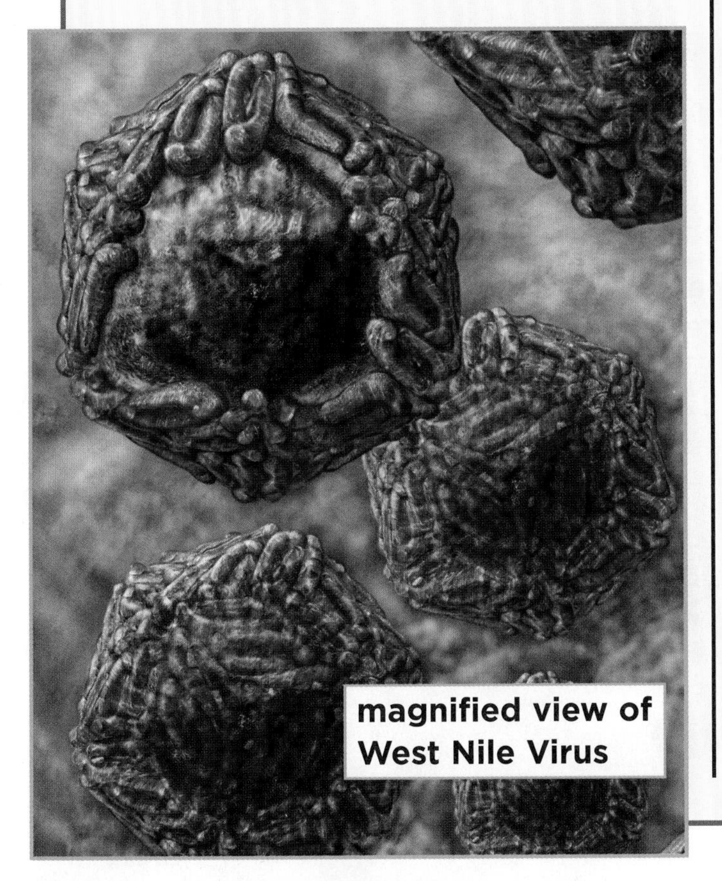

magnified view of West Nile Virus

3. **True or False** Write T if the statement is true or F if it is false.

   ___Biotechnology uses living things to meet human needs.

   ___Medical technology is an example of biotechnology.

   ___Lasers cut through skin without causing any bleeding.

   ___Medical technology does not usually depend on science.

   ___Biotechnology can have both good effects and bad effects.

   ___Chemicals can be used in biotechnology.

   ___Medical technology can help doctors see inside the body.

4. **Writing Link** Think about the different uses of medical technology you read about in this lesson. Which use do you think is most important? Why? Make a drawing showing how this technology might be used. Write a short paragraph that explains your ideas.

-Tech Link For more on medical technology and biotechnology go to **www.macmillanmh.com**

Lesson 5

Main Idea

Technology makes industry more productive.

Vocabulary

industry, p. 32

manufacturing, p. 32

consumer, p. 32

profit, p. 33

mass production, p. 36

productivity, p. 36

robot, p. 37

▲ designer's sketch

Technology in Industry

Are you familiar with the word "industry?" When you hear it, do you think of factories? Well, industry includes factories, but much more, too. **Industry** is all the work people do, from manufacturing goods to providing services.

Look at the clothes you're wearing. People working in the textile industry designed and manufactured your clothing. **Manufacturing** is how raw materials, such as cloth (or fabric), are made into goods, such as clothing. Goods are sold to consumers. A **consumer** is someone who buys things.

There are many different kinds of industries. Look around you. All the products you see were made by one kind of industry or another.

▼ **A woman works on an assembly line in the textile industry.**

Some industries rely on other industries. The computer industry makes computer systems and software. But computers are also important in almost every other industry. For example, they can store data that helps company owners figure out how much profit they'll make. It costs money to make products. A **profit** is the money a company makes after paying these costs.

The entertainment industry includes sports, music, theater, TV, and movies.

Computers are also used to make presentations. Using special software, people input words, graphs, and photos about the company and its products. Then they display it at industry meetings.

Students use computers to do research and write reports.

Quick Check

Compare and Contrast How are the industries you have read about similar? How are they different?

Workers and Industry

People working in the construction industry build things. Companies hire people to do specific tasks. Let's see what goes into building a school.

1 Planning Stage

First, people in a community decide they need a new building where kids can learn. They work with bankers to get money to pay for the construction. They find a site to build on. Architects and engineers work together to make blueprints for the building. They have to follow building codes, or rules. One code might say the building has to have a certain number of doors in case there is an emergency.

2 Laying a Foundation

Construction workers use cranes, bulldozers, and tractors to dig out an area. They fill the area with concrete to make the building's foundation. A foundation is the base that gives support to the building. Support columns made of steel and concrete are built into the foundation. So are pipes and tunnels for water and sewage systems.

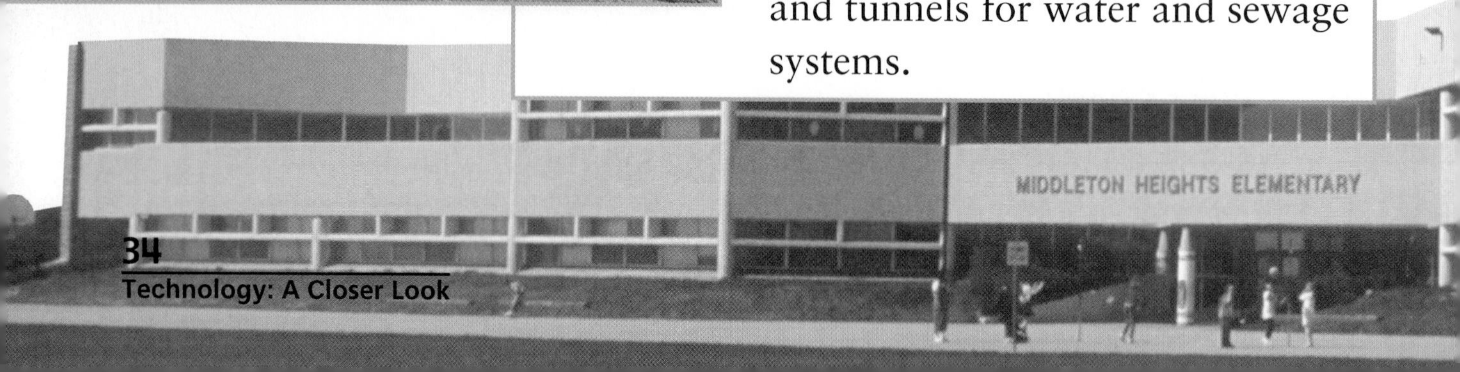

3 Building Stage

Construction is a system in which the workers and the materials are the input. Machines and methods make up the process. The output is the finished building.

During construction, workers build a frame of the building. They use materials like steel, concrete, and wood. Plumbers build in a system to bring water into the building and take waste out. Heating, ventilation, and air conditioning are built in, too. Electricians wire the building for power. Then workers cover the walls with dry wall and plaster.

4 Finishing Stage

After the frame is built and various systems are put in place, workers finish the school. They paint the walls and put down flooring. They put in windows, doors, and ceilings. Next time you're at school, look around. Everything you see was put there by someone during the construction process.

Quick Check

Compare and Contrast How are the different workers involved in the construction industry alike? How are they different?

Mass Production

To build a school takes a lot of materials. Materials used for construction are manufactured, just like the goods you buy at the store.

Manufacturing uses **mass production**. Mass production means that goods are built piece-by-piece on an assembly line. The assembly line was Henry Ford's idea. He built it to mass-produce cars.

On Ford's assembly line, cars went from one worker to another. Each worker had a specific job to do to each car as it came by. They increased **productivity**, or did more work in less time. That let Ford save money, so he was able to sell his cars for less.

Manufacturing Paint

In this factory, paint is made in huge vats (left). Then workers on an assembly line put the paint into cans (above). The cans of paint are then shipped to stores for sale to consumers.

Today many industries use robots to improve productivity. A **robot** is a machine that can repeat the same task over and over very precisely. In factories, robots often perform tasks too boring or dangerous for humans. They might mix chemicals, spray-paint items, or package products. Robots don't get bored or tired.

Quick Check

Compare and Contrast Draw a Venn diagram to show how humans and robots are alike and different.

Read a Diagram

What stages in paint-making do the photos show?

Clue: Consider what you know about input, process, and output.

Tech Activity

Divide the Work

Materials **paper, markers or crayons, timer**

What to Do

1. Draw a simple building like the one below.

2. Copy your drawing as many times as you can in 3 minutes. Use a clock or timer.
3. How many drawings did you complete?
4. Break the work into four tasks: walls, roof, door, and window.
5. Give each person in the group a specific task. The drawing should be passed from person to person as each task has been completed.
6. Complete as many drawings as possible in 3 minutes.
7. Count the number of drawings your group made. Is it more, the same, or less than you made alone?

Positively Plastic!

What comes in every color and is in almost every home in the United States? Something made of plastic! We sit on plastic chairs, play with plastic toys, and drink from plastic cups.

Plastics aren't found in nature. They are man-made from petroleum oil. Most plastics start out as thick, black oil. Plastics are strong, lightweight, and durable. There are many different kinds of plastics. They have replaced metals in making many products, such as car parts. Polyester is a kind of soft plastic used in making clothing.

The first modern plastics were developed in the 1930s. Today manufacturing plastics is a major industry. The plastics industry supplies materials for everything from TVs to cell phones to computers to artificial body parts. Some kinds of plastics can even be burned for energy. Plastics are useful, but they've also created a problem for the environment.

Remember, plastics are made from oil, one of Earth's nonrenewable natural resources. When we use up all the oil, there's no more. Remember also that plastics last a long time. The plastics buried in a garbage dump will be there for thousands of years. People have started to recycle plastics. Recycling is important because it saves natural resources and cuts down on garbage.

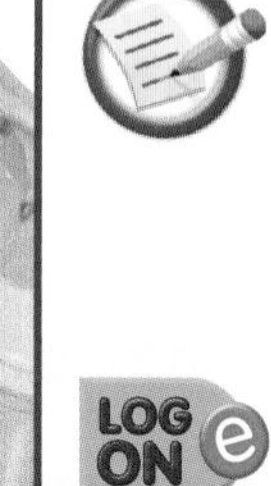

Write About It

Cause and Effect This article is about plastics. Research more about how plastics are made or recycled. Then write a report. Use facts and details from the article and your research.

LOG ON e-Journal Research and write about it online at www.macmillanmh.com

Lesson Review

Think, Talk, and Write

Write your answers on a separate piece of paper.

1. **Main Idea** What is industry? Write your own definition and provide some examples.

2. **Critical Thinking** Explain how profit and productivity are related.

3. **Compare and Contrast** How is the construction industry similar to the computer industry? How are they different?

4. **Fill in the Blank**

   A Industry involves manufacturing ________ and providing ________.

   B Making a lot of the same thing at the same time is called ________.

   C Robots help if a job is too boring or too ________ for humans.

   D Plastic is made from ________.

Word Box	
dangerous	mass production
services	goods
petroleum oil	

5. **Think About It** List the different parts of the construction industry that helped make the school pictured below. Use what you read about in the lesson, but don't stop there. For example, think about how the building materials were made. How were they shipped to the construction site? Next, work with a partner and list parts of the entertainment industry.

6. **Writing Link** Factory workers can use a computer program to change a mass-produced product to fit each customer's needs. Do you think this is important? Why or why not? Write a short paragraph to explain.

LOG ON e-Tech Link For more on technology in industry go to www.macmillanmh.com

Lesson 6

Main Idea
Technology can improve our lives, but it can also cause problems.

Vocabulary

society, p. 40
impact, p. 42
environment, p. 42
law, p. 42

Technology, Society, and the Environment

You've learned about the difference modern technology can make in our lives. When we use the term "our lives" here, what we mean is modern society. A **society** is a group of people living as members of a community.

Early Technology

One of the earliest societies lived during the Stone Age. During this period, people discovered how to make tools, weapons, bridges, and monuments out of stone.

Copper and tin are weak metals. Mixed together, they form bronze—a strong metal. During the Bronze Age people used this material to make tools and weapons for hunting.

Next came the Iron Age, when iron replaced bronze as the basic material for making tools and weapons.

▼ **The ancient Romans, who lived more than 2,000 years ago, built structures from stone.**

▲ Today materials include glass, concrete and steel.

Modern Technology

The Stone, Bronze and Iron Ages paved the way for later societies. Eventually, people learned how to make steel and concrete to build buildings and roads. These discoveries led to the Industrial Revolution, a time of rapid advances in technology.

The Industrial Revolution began in the 17th Century, with inventions like the steam engine. The steam engine allowed people to make goods in manufacturing plants, or factories. People who had been farmers went to work in factories. They moved to homes near the factories. Soon cities grew up. People's lives changed.

The Space Age began in 1957, with the launch of the first artificial satellite. The Nuclear Age soon followed, with the splitting of the atom to make electricity.

Technology continues to change our lives. Today, we are in an age of computers. Some call it the "Cybernetic (sy•bur•NET•ik) Age." New inventions and discoveries take place every day.

There is one thing that links past, present, and future societies. Each society advanced when someone made an interesting discovery. What do you think will be discovered next? Travel to other galaxies? Time travel? Maybe you will make a discovery!

Quick Check

Draw Conclusions How does the invention of new technology change a society?

Technology and Nature

As we have learned, technology always impacts a society. An **impact** is an effect, or consequence, of something. Technology can also impact the **environment**. The environment includes all the living things in an area and their surroundings. The way we use technology can have good or bad effects on the environment.

Consider the Trans-Alaska Pipeline System. It was built in the 1970s to move oil across Alaska. Oil is a thick, dark liquid. People value oil because it's used for fuels, like gasoline, and to make materials, like plastics. The pipeline system is about 1,300 kilometers (800 miles) long. It crosses three mountain ranges and more than 800 rivers and streams. Before building the pipeline, the government had to think about its impact on nature. It had to make sure that the pipeline was strong, because if oil spilled from a break, it would harm water, forests, and animals.

When the Alaska pipeline was built, there were laws protecting the environment. A **law** is a rule that tells what people can and cannot do. One law said the pipeline couldn't be built if it harmed the environment. Another required a plan to protect the environment if there were an accident.

Living with Technology

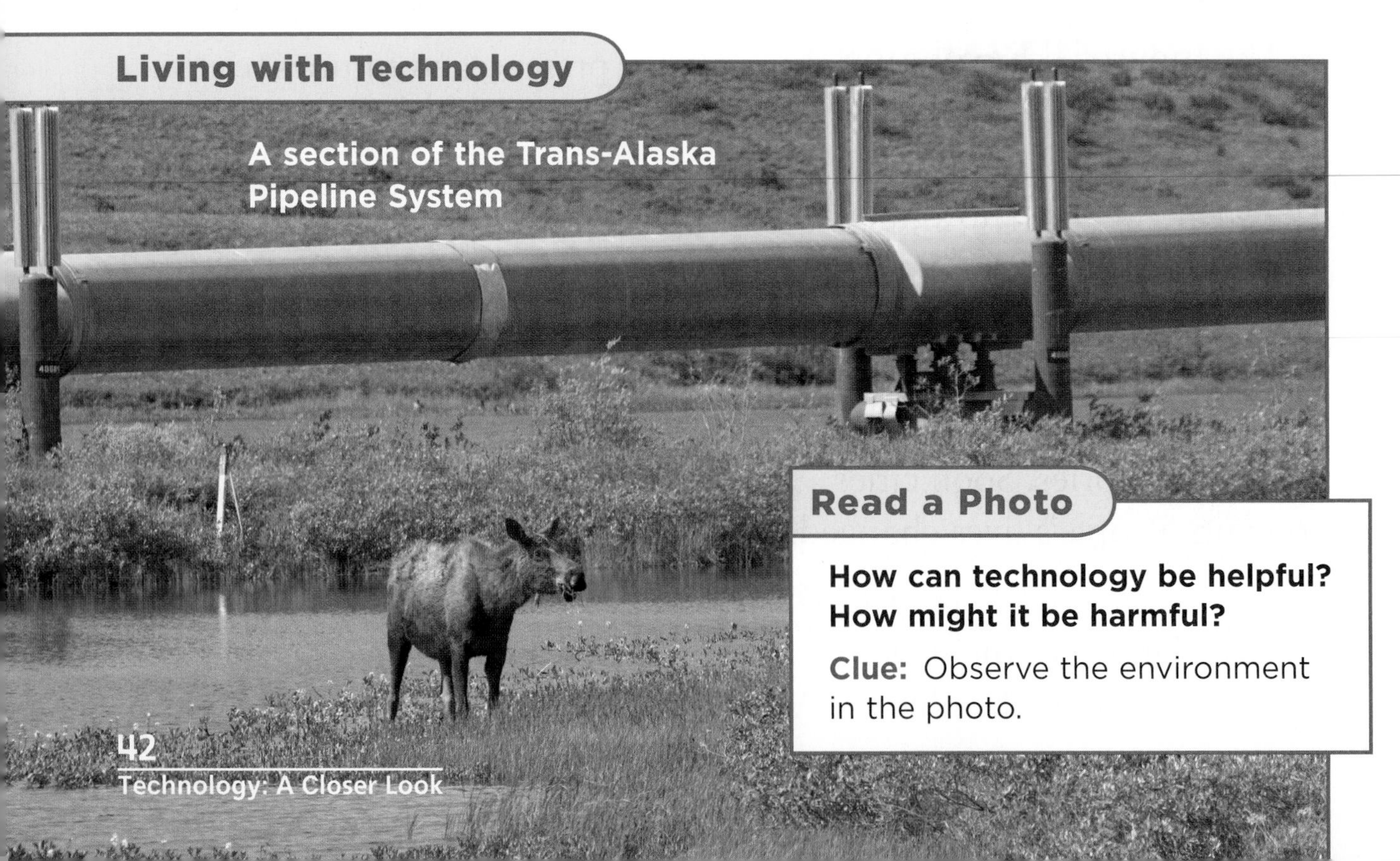

A section of the Trans-Alaska Pipeline System

Read a Photo

How can technology be helpful? How might it be harmful?

Clue: Observe the environment in the photo.

Following the Rules

Before the 1970s, there weren't many laws protecting the environment. Companies dumped unused chemicals into rivers and streams. Builders used asbestos behind walls to help keep homes warm. They used lead paint.

Back then, people didn't think about how chemicals would affect drinking water. They didn't realize that asbestos could cause lung disease or that lead paint could damage a young child's brain.

Today there are many laws to protect society and nature. These include laws that stopped companies from dumping chemicals into rivers. Builders aren't allowed to use asbestos. Manufacturers can't put lead in paint.

Laws help protect our water, food, and air.

Quick Check

Draw Conclusions Why is it important to consider the impact of technology on the environment?

Tech Activity

Make a Sun Dial

For more than 2,000 years the sun dial was the most common timekeeper in use.

Materials **sun dial rectangle, a piece of thin cardboard, glue, hole punch, a piece of string, tape, scissors**

What to Do

1. △ **Be Careful.** Cut around the outside edge of the sun dial. Glue the rectangle to the cardboard. Fold along the center line.
2. Punch a hole in the center of each end of the sun dial. Thread the string through the two holes as shown, taping the ends to the back of the cardboard.
3. **Observe** On a sunny day, hold the card level and point it toward the north. Observe the shadow of the string on the sun dial. Which hour mark is it near? What time does your sun dial give?

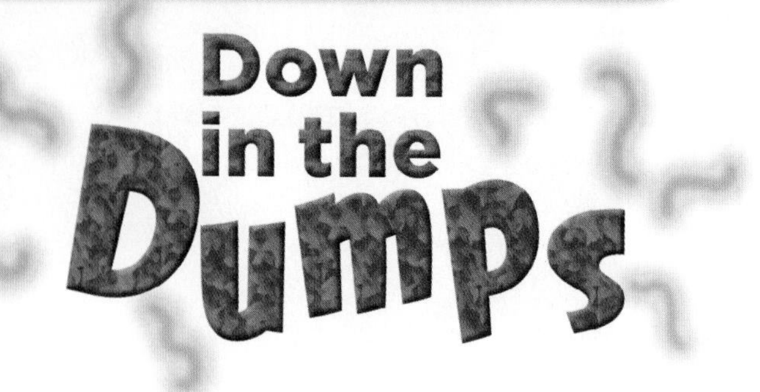

Did you know the average American throws away about five pounds of trash each day? For years, we “got rid” of trash by burying it in dumps or landfills. The idea was to cover trash with dirt and let bacteria go to work. The bacteria would decompose the garbage and make the soil richer. Good idea, right?

Wrong! Scientists dug into old landfills to observe the effect of bacteria. They found newspapers from the 1970s that were still readable. They found green grass clippings and a T-bone steak with fat on it. They also found five hot dogs that still looked like hot dogs! The garbage hadn’t decomposed. Scientists experimented to find out why. It turns out that most landfills don’t get enough air and moisture for bacteria to grow.

Landfills today are very different from the ones in the past. New technology helps turn trash into fuel. Landfills give off methane gas. If you seal the landfill, the gas builds up. Then scientists collect the gas and send it through pipelines to provide needed energy.

Write About It

Persuasive Writing Write a letter persuading your mayor to make sure landfills in your area help conserve Earth’s natural resources. Use facts and details from the article in your letter.

LOG ON e-Journal Research and write about it online at www.macmillanmh.com

Lesson Review

Think, Talk, and Write

Write your answers on a separate sheet of paper.

1. **Main Idea** Discuss some ways that technology impacts society and the environment.

2. **Compare and Contrast** Describe some ways the technology of the Stone Age differed from today's technology.

3. **Critical Thinking** Explain the things that people do to cut down on the trash they make.

4. **Complete the Sentences** Choose your answer from the following words.

Word Box	
impact	environment
society	laws
landfill	

A A ________ is an area where people dump garbage.

B The effect of one thing on another is also called an ________.

C The ________ includes living things and their surroundings.

D Another word for a group of people that live together in a community is ________.

E Before the 1970s, we did not have many ________ to protect the environment.

5. **Writing Link** Think about automobiles. What positive effects do they have on society? What negative effects? Write a paragraph to explain.

e-Tech Link For more on technology's impact on society and the environment go to www.macmillanmh.com

Lesson 7

Main Idea
Technology will continue to impact society and shape the future.

Vocabulary

Technology and the Future

What do you plan to do after school today? What will you be doing at the same time five years from now? Twenty years from now?

It's impossible to know for sure what the future will be like, but we can make predictions. To **predict** means to use what you know to make a guess about the future.

How do scientists predict the future? They study the past. Then they look at what scientists know today. This information helps them come up with ideas about the future.

These paintings show how two artists think the future will look.

The Internet

In the future, the Internet will continue to improve the ways we work and communicate. Today, you can send information around the world in seconds.

Still, scientists are working on an even faster Internet called Internet2. You'll be able to use it in ways that aren't possible on today's Internet. You'll program and control robots with it. Doctors will be able to use it for surgery!

Computers and robots like this lawn mower may play an important role in the future.

Globalization

Faster communication, working across long distances—these are paths to **globalization** (glow•buhl•ih•ZAY•shuhn). Globalization is the way technology makes the world seem like a smaller place. It helps people around the world work with one another in business. One day you may live in the United States but work for a company in another country using Internet2.

Technology isn't available to everyone who needs it, such as people in poor countries. Making technology available worldwide is a problem to solve.

Quick Check

Infer What kinds of information might help you make a prediction about the future?

Energy to Burn

Another problem to solve for the future is energy. Most of today's energy comes from fossil fuels—oil, coal, and natural gas. Fossil fuels are not easy to replace in nature. Someday they will be used up. Burning fossil fuels causes pollution.

People are using technology to find new and better sources of energy. We already know how to get energy from the wind, water, and Sun. Some buildings, for instance, have solar panels to collect the Sun's energy. In the future, these technologies might deliver more of our electricity.

Biomass Energy

What will the cars of tomorrow be like? We already have hybrid cars powered by electricity and gas. Could cars run on what living things leave behind? They just might!

This minicar can run on ethanol.

You may think of it as garbage, but **biomass** can be used for energy. Biomass is matter from living things, such as dead leaves and twigs. These can be dried then burned to make electricity, heat, and fuels.

One example of biomass energy is ethanol (ETH•uh•nawl). This fuel is also called grain alcohol. Ethanol can be made from grass clippings, cow manure, and other forms of biomass. People are working on technology that uses ethanol to power cars.

Cassava root (left) and oil palm fruit (above) are used to produce biofuels or fuels made from biomass.

Nanotechnology

Scientists predict that nanotechnology will be important in the future. Nanotechnology uses really, really small materials to make products.

Today nanotechnology is used to make sunglasses that block the Sun and don't scratch easily. It is used to make materials to coat fabrics with, to prevent stains and wrinkles. Some sunscreens have "nanoparticles" to help protect people from the Sun's harmful rays Scientists are working on more ways to use nanotechnology.

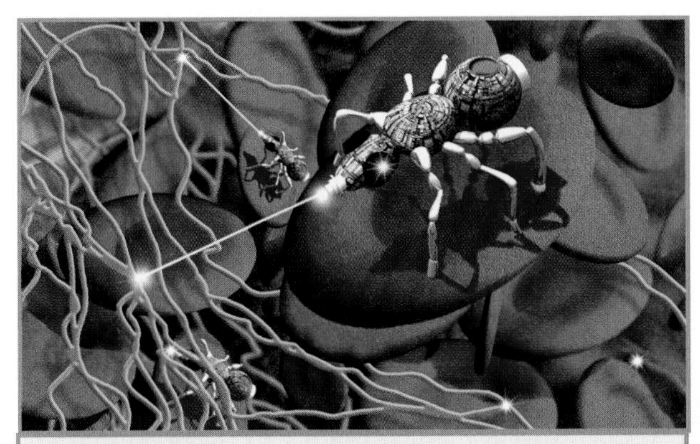

Some day doctors may inject you with "nanobots." These tiny robots would find and destroy diseases.

Tech Activity

Is it Stronger?

Some scientists want to make stronger buildings by weaving strings of "nanotubes". Find out why!

Materials **two or three sheets of 8″ x 11″ poster board, several strips of 1″ x 11″ poster board, ruler pencil, scissors**

1. **Predict** Does weaving affect the strength of a material?
2. Fold a sheet of poster board in half. Draw lines one-inch apart from the fold to about one inch from the outer edge. Carefully cut along each line.
3. Lay the poster board on a flat surface. Weave the strips over and under the cuts.

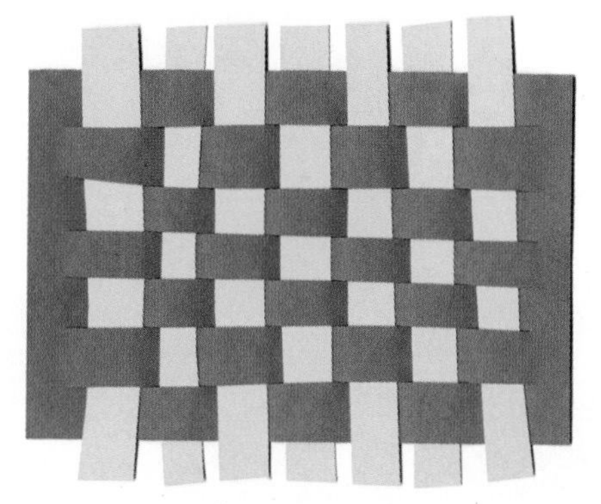

4. **Observe** Rip the woven material. Rip a sheet of plain poster board. Compare what happened.
5. Check your prediction.

Quick Check

Infer Some people call nanotechnology "the new Industrial Revolution." What does this mean? Why might people think so?

Go for the GLOW!

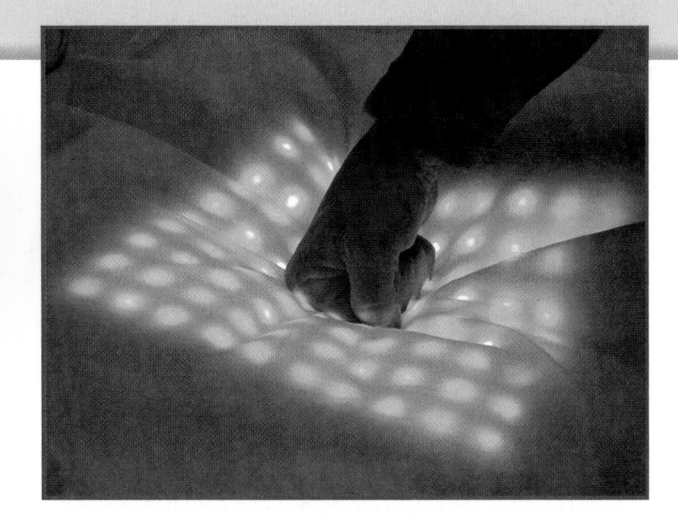

Shirts that light up? Animated cartoons on your bedsheets?

Scientists are testing nanotechnology on fabrics by weaving colored "nanolights" into them. Nanolights are actually tiny light-emitting diodes (DY•odz), or LEDs. Their small size helps the fabric stay soft and bendable.

With such fabrics, it would be possible to make clothing that flashes words or changes color. The fabrics might display moving pictures or the time of day! The fabrics will also be used for making drapes, curtains, and furniture coverings.

These light-emitting fabrics have other uses. Companies might use them to advertise goods or services. Guides at museums and science centers might wear them to be more visible. What other uses can you think of?

Write About It

Fictional Narrative This article tells about a kind of nanotechnology used in fabrics. Review what you've read in the article. Then write a fictional story about a kind of nanotechnology in your life 20 years from now.

LOG ON e-Journal Research and write about it online at **www.macmillanmh.com**

Lesson Review

Think, Talk, and Write

Write your answers on a separate sheet of paper.

1. **Main Idea** How do scientists make predictions about the future? What do they base their predictions on?

2. **Use Numbers** The word nanotechnology comes from the word "nanometer"—a unit of measure. A nanometer is one-billionth of a meter. A billion is 1,000 times bigger than a million. You can write one million as 1,000,000. How would you write one billion as a number?

3. **Critical Thinking** Choose an everyday object that you'd like to make smaller, stronger, or better with nanotechnology. If you were a scientist, where would you begin?

4. **Complete the Sentences** Choose your answers from the following words.

Word Box	
small	energy
globalization	twigs
predict	nanolights
leaves	products

A To ________ means to make a guess about the future.

B ________ is the way technology makes the world seem like a smaller place.

C Biomass, such as dead ________ and ________, can be used for ________.

D Nanotechnolgy uses really ________ materials to make ________.

E Scientists have developed a way of weaving ________ into fabrics.

5. **Draw and Write On!** What do you think a family car will look like in the year 2055? Draw your idea of a future car in a magazine ad that tells why people should buy it and how much it will cost.

-Tech Link For more on technology of the future go to www.macmillanmh.com

Glossary

Pronunciation Key

The following symbols are used throughout this glossary.

a	at	e	end	o	hot	u	up	hw	white	ə	about
ā	ape	ē	ma	ō	old	ū	use	ng	song		taken
ä	far	i	it	ôr	fork	ü	rule	th	thin		pencil
âr	care	ī	ice	oi	oil	u̇	pull	th	this		lemon
ô	law	îr	pierce	ou	out	ûr	turn	zh	measure		circus

′ = primary accent; shows which syllable takes the main stress, such as **kil** in **kilogram** (kil**′** e gram′).
′ = secondary accent; shows which syllables take lighter stresses, such as **gram** in **kilogram**.

B

biomass (bī′ō·mas′) Plant matter and animal wastes that can be used as a source of fuel. (p. 48)

biotechnology (bī′ō·tek·nol′əjē) The use of living things to make products that improve human life. (p. 29)

C

communicate (kəmū′ni·cāt) To share your ideas with others. (p. 16)

consumer (kən·sü·′mər) A person who buys goods or services. (p. 32)

D

design process (di·zīn′ pros′es) The process through which an idea becomes an invention or technological solution. (p. 9)

E

environment (en·vī′rən·mənt) All the living and nonliving things in an area. (p. 42)

G

globalization (glōb′li·záysh′n) The process of making goods and services available to the other countries of the world. (p. 47)

goods (gōōdz) Things we make, grow, buy, and sell such as food and clothes. (p. 4)

I

impact (im′pakt) A result or effect, often unintended. (p. 42)

industry (in′dəstrē) A group of businesses that provide similar goods or services. (p. 32)

L

law (law) A rule that tells what people can and cannot do. (p. 42)

M

manufacturing (mànnyəfákchər·ing) The process of changing raw materials into finished products. (p. 32)

mass production (mas prəduk'shən) The manufacturing of a large volume of products very quickly. (p. 36)

model (mod'əl) A representation of something that cannot be directly observed. (p. 12)

N

nanotechnology (nan'ətek·nol'əjē) Using tiny particles to create or improve goods and services. (p. 49)

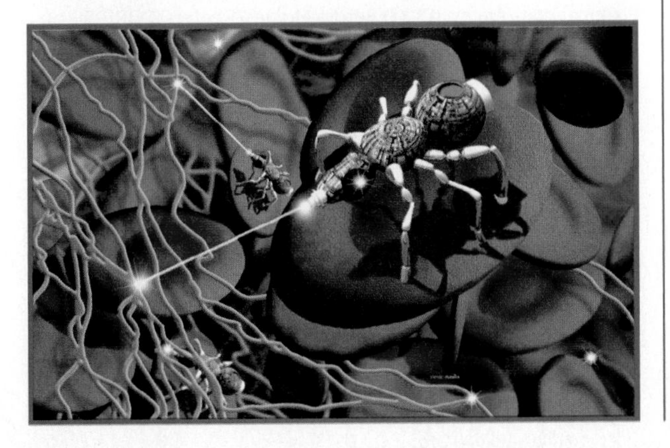

natural resource (nach'ər·əl rē'sôrs') A material on or in the Earth that is useful to people. (p. 2)

P

patent (pátt'nt) A set of rights for the inventor or manufacturer of a product. (p. 9)

productivity (prò·duk·tívvətee) A measurement of the goods and services made over a given amount of time. (p. 36)

predict (pri dikt') To state likely results based on what you know or observe. (p. 46)

profit (próffit) The amount of money collected from goods or services after the cost of producing them is subtracted. (p. 33)

prototype (prō'tətīp') A full-sized, working model of a product that is usually tested. (p. 12)

R

robot (rō'bot) A machine that automatically performs a task over and over again. (p. 37)

S

service (súrvəss) Work done by a person or a group that benefits another person or group. (p. 4)

society (səsí·ətee) A group of people with a common interest. (p. 40)

system (sis'təm) A group of separate parts that work together to do something. (p. 3)

T

technology (tech·nol'əjē) All the ideas and tools humans design, make, and use to fill their needs. (p. 2)

telegraph (téllәgràf) A communication device that sends messages over long distances. (p. 16)

trait (trāt) A characteristic of a living thing, such as eye color. (p. 29)

Credits

Illustration Credits: 2-3: Robert Roper. 10, 12, 13, 15: Jackie Stafford-Snider. 18, 19, 23: Christine Schneider. 45: Jackie Stafford-Snider.

All Photography Credits: All photographs are by Macmillan/McGraw-Hill (MMH) except as noted below:

ii: (t) Photodisc/Getty Images; (c) POPPERFOTO/Alamy; (b) Stockbyte/PunchStock. iii: (tr) Victor Habbick Visions/Photo Researchers, Inc.; (tl) Science Museum/SSPL/The Image Works; (c) SSPL/The Image Works; (b) Arcaid/Alamy. 2: (l) Science Museum/SSPL/The Image Works. 3: (t) Bettmann/CORBIS. 4: (t) Kean Collection/Getty Images; (l) Hulton-Deutsch Col lection/CORBIS; 4-5: (r) Danny Lehman/CORBIS. 5: (b) Richard Levine/Alamy. 6: Myrleen Ferguson Cate / PhotoEdit. 7: (t) Dynamic Graphics Group/Creatas/Alamy; (b) Victor Habbick Visions/Photo Researchers, Inc. 8: (l) SSPL/The Image Works; (r) Topham/The Image Works. 9: (c) Information Technology and Society Division, NMAH/Smithsonian Institution; (b) PhotoSpin/Alamy. 14: (t) POPPERFOTO/Alamy; (b) Philippe Plailly/Photo Researchers, Inc. 15: (t) PhotoSpin/Alamy. 16: (r) Tetra Images/Alamy; (l) Bettmann/CORBIS. 17: Image100/PunchStock. 18: (t) Andre Jenny/Alamy. 19: (t) Photodisc/Punchstock. 20: (c) D. Hurst/Alamy; (b) David J. Green/Alamy; (inset) The Granger Collection, New York; (t) Image Farm; (inset) Photolink/Getty Images. 21: (t) Bettmann/CORBIS; (b) Walt Disney Co./Courtesy Everett Collection. 22: (b) StockTrek/Getty Images; (t) Brand X Pictures/PunchStock; (l) Photodisc/PunchStock; (r) Brand X Pictures/PunchStock; (inset) POPPERFOTO/Alamy. 24: (l) Neil Fletcher and Matthew Ward/Dorling Kindersley; (r) Custom Medical Stock Photo. 25: (br) Brand X Pictures/Alamy; (bl) Jim Cummins/Getty Images; (tl) Ulrich Baumgarten/Vario Images BmbH & Co.KG/Alamy; (tr) Chris Jones/CORBIS. 26: (b) National Cancer Institute/Getty Images; (t) Photodisc/Getty Images. 27: (b) Medical-on-Line/Alamy; (t) Richard T. Nowitz/Photo Researchers. 28: Ralph Wetmore/Alamy. 29: (t) Meul/ARCO/naturepl.com; (b) Wolfgang Flamisch/zefa/Corbis. 30: Richard Hutchings/CORBIS. 31: Russell Kightley/Photo Researchers, Inc. 32: ullstein-Wodicka/Peter Arnold, Inc. 33: (t) DigitalVision/PunchStock; (b) Rob Crandall/The Stock Connection. 34: (c) ImageState/Alamy; (bl) Robert W. Ginn/PhotoEdit; (t) prettyfoto/Alamy; 34-35: (b) David R. Frazier Photolibrary, Inc./Alamy. 35: (t) Peter Titmuss/Alamy; (c) Kathy McLaughlin/The Image Works. 36: (l) Rosenfeld Images Ltd./Photo Researchers, Inc.; (r) John Sturrock/Alamy. 37: Stockbyte/Punchstock. 38: (b) Comstock/Alamy; (c) Stockbyte/PunchStock; (t) Image Source/PunchStock. 39: David Frazier/PhotoEdit. 40: John W Banagan/Getty Images. 41: Arcaid/Alamy. 42: Alaskastock. 43: Charlie Rahm/USDA Natural Resources Conservation Service. 44: Reuters/CORBIS. 46: (l) Forrest J. Ackerman Collection/CORBIS; 46-47: (r) Fujifotos/The Image Works. 47: Bo Jansson/Pixonnet.com/Alamy. 48: (t) Sergio Moraes/Reuters/CORBIS; (bl) John Van Hasselt/Corbis; (br) John Van Hasselt/Corbis. 49: (l) Victor Habbick Visions/Photo Researchers, Inc. 50: (b) Philips; (t) Philips. 51: Iain Masterton/Alamy. 52: Wolfgang Flamisch/zefa/Corbis. 53: (t) ullstein-Wodicka/Peter Arnold, Inc.; (b) Victor Habbick Visions/Photo Researchers, Inc. 54: (inset) Photolink/Getty Images; (t) David J. Green/Alamy.